Ahmed Abd EL-kader

Targeting Cytoplasmic Tyrosine Kinases in Cancer:

AF791005

Ahmed Abd EL-kader

Targeting Cytoplasmic Tyrosine Kinases in Cancer:

The Paradigm of Bcr-Abl in Chronic Myeloid Leukemia

Noor Publishing

Imprint
Any brand names and product names mentioned in this book are subject to trademark, brand or patent protection and are trademarks or registered trademarks of their respective holders. The use of brand names, product names, common names, trade names, product descriptions etc. even without a particular marking in this work is in no way to be construed to mean that such names may be regarded as unrestricted in respect of trademark and brand protection legislation and could thus be used by anyone.

Cover image: www.ingimage.com

Publisher:
Noor Publishing
is a trademark of
Dodo Books Indian Ocean Ltd. and OmniScriptum S.R.L publishing group

120 High Road, East Finchley, London, N2 9ED, United Kingdom
Str. Armeneasca 28/1, office 1, Chisinau MD-2012, Republic of Moldova, Europe
Managing Directors: Ieva Konstantinova, Victoria Ursu
info@omniscriptum.com

Printed at: see last page
ISBN: 978-613-8-39249-1

Targeting Cytoplasmic Tyrosine Kinases in Cancer: The Paradigm of Bcr-Abl in Chronic Myeloid Leukemia

Ahmed Abdelkader

Department of Pharmaceutical Chemistry, Collage of Pharmacy, Al-Farahidi University ,Baghdad, 10021, Iraq

Corresponding Author

Dr : Ahmed Abdelkader

Department of Pharmaceutical Chemistry, Collage of Pharmacy, Al-Farahidi University ,Baghdad, 10021, Iraq

Email : aabdelkader75@yahoo.com, ahmed.abdelkader@uoalfarahidi.edu.iq

Table of content

Abstract

Protein kinases are critical regulators of cell signaling and represent one of the most successful classes of drug targets in oncology. Among them, cytoplasmic tyrosine kinases (CTKs) play a pivotal role in controlling cell growth, survival, and differentiation. Their dysregulation through mutation or translocation is a common driver of oncogenesis. This review provides a comprehensive overview of CTKs as therapeutic targets, with a particular focus on the c-Abl kinase and its oncogenic counterpart, Bcr-Abl, in Chronic Myeloid Leukemia (CML). We detail the structural biology of c-Abl, including its autoinhibitory mechanism, and explain how its fusion with Bcr leads to constitutive activation, driving leukemogenesis. A significant portion of the review is dedicated to the development and clinical evolution of Bcr-Abl inhibitors, from first-generation ATP-competitive drugs like imatinib to second- and third-generation agents designed to overcome resistance. We further explore the promise of allosteric inhibitors, which bind to sites beyond the ATP-pocket, and highlight the synergistic potential of combination therapies that target both orthosteric and allosteric sites to suppress resistance mutations, including the recalcitrant T315I "gatekeeper" mutation. Finally, we discuss the clinical landscape of kinase inhibitors and future perspectives, positioning the Bcr-Abl story as a paradigm for targeted cancer therapy.

Key word s

Bcr-Abl , Chronic Myeloid Leukemia (CML) , Tyrosine Kinase Inhibitors

(TKIs) , Drug Resistance, Allosteric Inhibition

1. *Over view of cancer*:

In spite of advances in diagnosis and treatment of cancer, cancer remains the second leading cause of death in the western world. It has been shown that cancer can be transmitted within or between cells (Dhillon et al., 2007). Cancer is a complicated genetic disease that is caused mostly by environmental factors. Carcinogens (cancer-causing agents) almost everywhere including food, water, chemicals, and in air. (Anand *et al.*, 2008)

Until recently, standard treatments using endocrine therapy, radiation therapy, chemotherapy, and surgery have been the available options for cancer patients. This has improved survival in many types of solid tumor, but unfortunately, treatment-associated toxicity and emergence of resistance have been the major cause of mortality and morbidity (Abou-Jawde et al., 2003). Therefore, a new and more effective therapy is urgently needed to enhance patient outcomes.

Rapid scientific progress in recent years has enhanced our knowledge of cancer biology which has led to the discovery of several novel targets, which results in treatments with better efficiency, high selectivity and low toxicity (S and S, 2015). Among these novel targets are protein kinases. Protein kinases are a family of phosphate transferase that are important for phosphorylation reaction, which is one of the most significant classes of post- translational modifications of protein(Masterson et al., 2012). From a pharmacological view point , small molecule inhibitors that are highly selective for protein kinases results in the inhibition of kinase functions at any step of cellular development (Dar and Shokat, 2011).

1.1Protein kinases in cancer

Protein kinases (PKs) play an important role in controlling several biological processes in establishing the correct cell functionality, and constitute nearly 2% of human gene (Manning *et al.*, 2002). Protein phosphorylation is a key biochemical hallmark of metazoans mediated by transferring a γ-phosphate of ATP and covalently attaching it to serine, threonine ,tyrosine or histidine residues belonging to peptide or proteins (Hunter, 2012) (**Figure 2.1**). About a third of all proteins in human cells are phosphorylated by protein kinase (Cohen, 2002b) .A significant change on the structure of the substrate protein occurs upon phosphorylation by protein kinases (Johnson and Lewis, 2001; Kuriyan and Eisenberg, 2007). This phosphorylation consequentially results in cell proliferation, migration and survival (Tarrant and Cole, 2009). Hence, Protein kinases are considered as key regulators of cell biology that are monitoring a number of signaling pathways which are vital for cell growth and survival. However, the overexpression of these kinases have been implicated in the initiation, promotion, progression as well as the reoccurrence of cancer and other diseases such as diabetes, vascular disease, inflammatory, and neurological diseases(Kang, 2014).

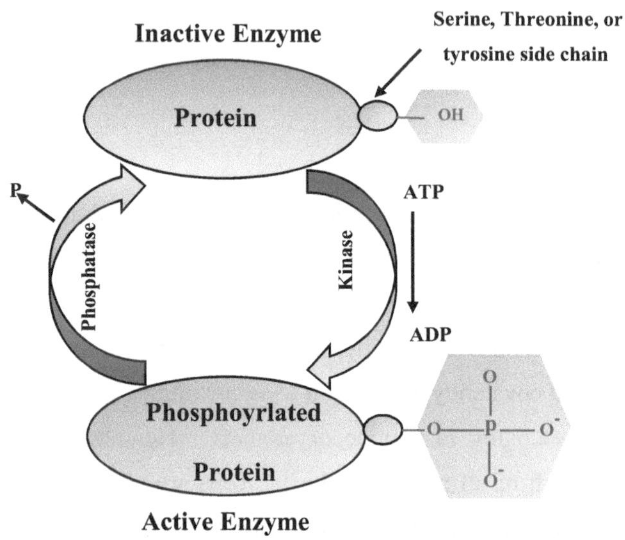

Figure 1: Protein phosphorylation pathways

This makes the inhibition or the modulation of protein kinases activity an attractive target in the treatment of several diseases especially cancer. Nearly 45 protein kinase inhibitors have been approved by US FDA, and thousands of clinical studies targeting kinases are ongoing(Kornev and Taylor, 2015; Wu, Nielsen and Clausen, 2015) **(Figure 2)**.

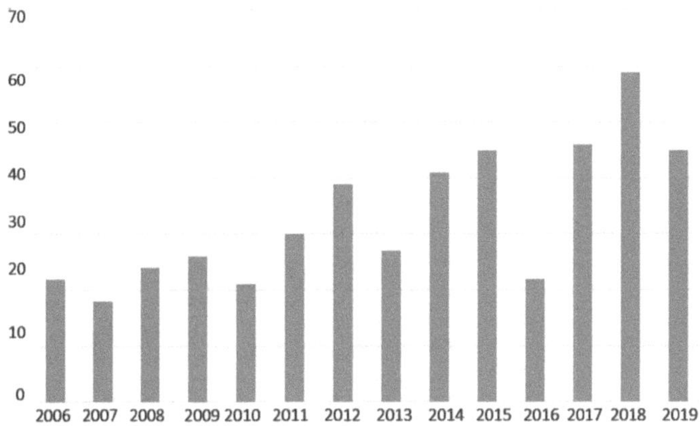

Figure 2: FDA approved protein kinase inhibitors .

1.2Classification of protein kinases

The human genome encodes 518 kinases from which about 90 are tyrosine kinases (Manning et al., 2002). The protein kinase family is divided into two types: cytoplasmic tyrosine kinases (CTKs), and transmembrane receptor tyrosine kinases (RTKs) which are further grouped into subfamilies based on protein sequence homology. The cytoplasmic tyrosine kinase (CTK) consist of 34 known members , all containing non-catalytic domains and a catalytic kinase domain (Robinson, Wu and Lin, 2000) **(Figure 3)**.

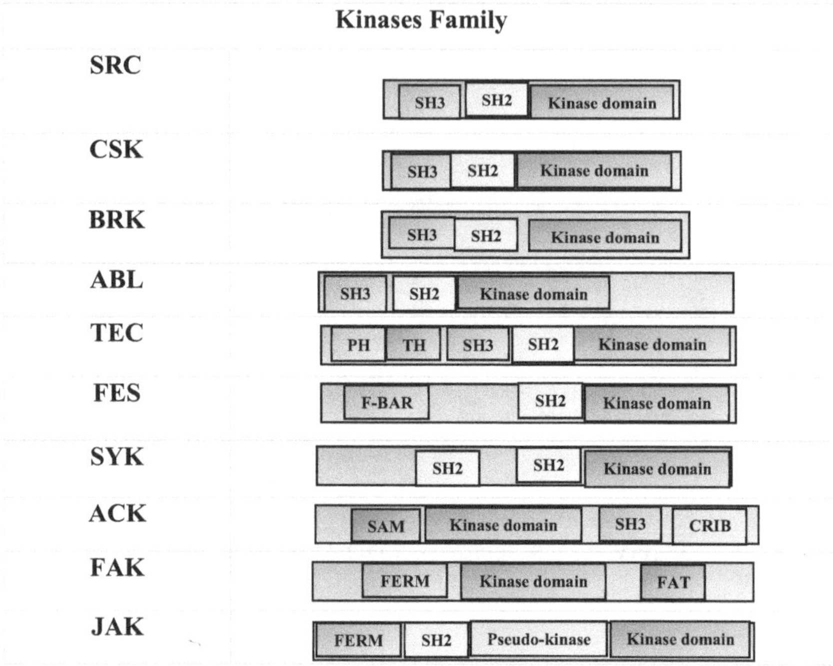

Figure 3: Structural organization of cytoplasmic tyrosine kinases Non-catalytic domain names are the following. **SH2/SH3**: Src-homology ; **PH**: Pleckstrin homology; **TH**: Tec-homology; **F-BAR**: Bin-Amphiphysin-Rvs; **SAM**: Sterile-α-motif; **CRIB**: Cdc42/Rac-interactive-binding-motif; **FERM**: protein-ezrin-radixin-moesin; **FAT**: focal adhesion- targeting

1.2.1 Non-catalytic domain

Non catalytic domain is of interest because of its high diversity as well as its significant role in cellular localization (Kung and Jura, 2016). For instance, non-catalytic domains such as SH2 which binds phospho-tyrosines allows spatial regulation and specific binding of CTKS to tyrosine-phosphrylated target proteins (Neet and Hunter, 1996). SH3 domains also play an important role in cellular substrate recognition, protein regulation and improved SH2-dependent phosphorylation (Pellicena and Miller, 2001). Other non-catalytic domains such as SAM, and CRIB are implicated in protein-protein interaction complexes. For example, the CRIB domain in ACK kinase plays an important role in its interaction with the activated form of Cdc42 GTPase to promote cell survival and multiplication (Kato-Stankiewicz et al., 2001). While, PH and FERM domains play a critical role on the localization of the membrane through binding to phospholipids which is significant for protein kinase activation pathways (Leonard and Hurley, 2011). Vihinen et al. has shown, the location of plasma membrane in Bruton tyrosine kinase (Btk) protein is an important step in B-cell activation, and the mutation of its PH domain can lead to X-linked agammaglobulinemia (XLA) (Vihinen et al., 1994).

1.2.2 Kinase domain

Kinase domain is highly conserved protein kinase domain that spans about 300 amino acids and catalyzes the transfer of phosphates to the hydroxyl group of protein substrates(Kornev *et al.*, 2006). The structure of the kinase domain can be subdivided into two domains N and C. The N domain consists of one prominent α helix and five stranded β-sheets, while the C domain is larger and mostly helical segment. (Cox, Radzio-Andzelm and

Taylor, 1994). ATP interacted with the deep cleft located between these two domain just below the conserved glycine rich loop. An optimal phosphate transfer requires a specific sequence of spatial arrangements of several residues located in the αC helix , the catalytic loop ,and the glycine rich loop (Huse and Kuriyan, 2002) (Figure 4). .

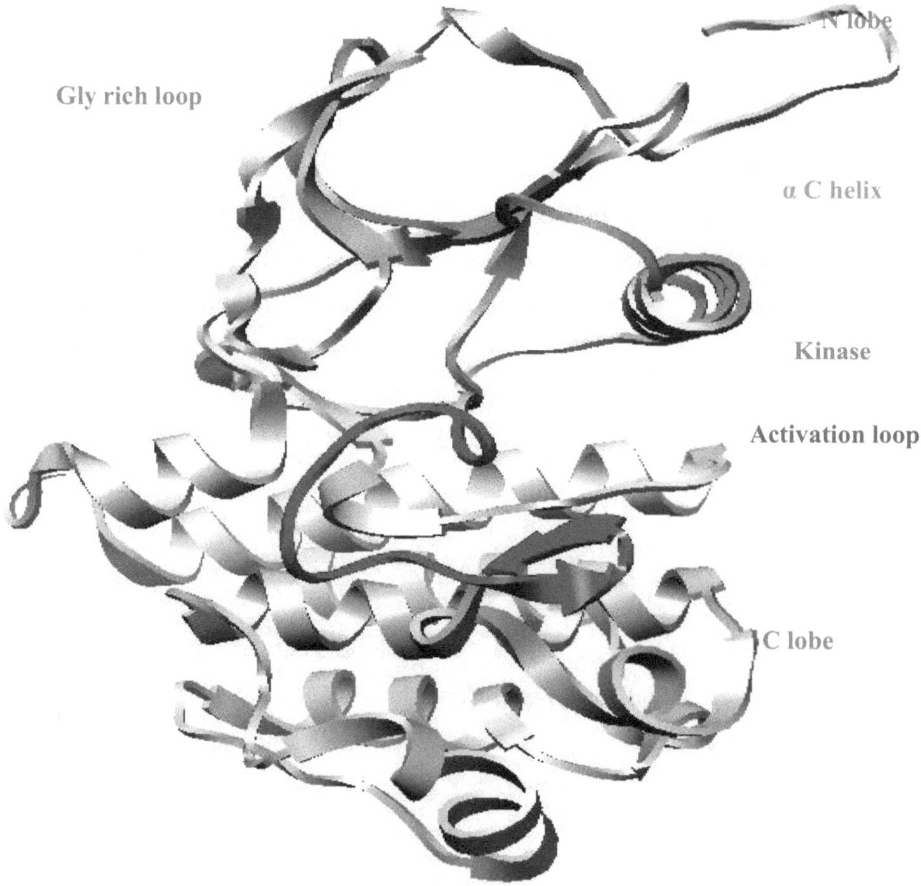

Figure 4: Cartoon representation of structural organization of the Abl kinase domain (PDB: 1IEP). N- and C- lobes are shown in gray. αC helix is shown in orange. The activation loop and the glycine rich loop are shown in red and blue respectively (Prepared by Author).

2.2.3 Rules of cytoplasmic tyrosine kinases in cancer

Cytoplasmic tyrosine kinase has a significant role in cellular signaling and deregulation by point mutation , translocation, deletions or duplications are usually combined by an over-activation that leads to cancer (Hubbard and Till, 2000; Blume-Jensen and Hunter, 2001). In 1978, Levinson et al discovered that there is a correlation between Src-family kinases (SFKs) and tumor progression of malignancy such as breast, colorectal, prostate or lung cancer (Zhang and Yu, 2012) .

1.3c- Abelson (Abl) kinase: a prototypic cytoplasmic tyrosine kinase

1.3.1 The Abl kinase family

c-Abl is a member of the cytoplasmic tyrosine kinase which is localized at several sites in the cell including nucleus, cytoplasm, mitochondria and endoplasmic reticulum (Wetzler *et al.*, 1993) . c-Abl is the prototype of a subfamily which includes two members : c-Abl (Abl1) and its paralogue Arg (Abl2,Abl related gene)(Sirvent, Benistant and Roche, 2008). Members of this family are highly conserved throughout the metazoans and ubiquitously expressed. Abl play a significant role in signaling pathways that is critical for cellular function such as adhesions, differentiation, division and stress response(Sirvent, Benistant and Roche, 2008).

1.4 Chronic myeloid leukemia and Bcr-Abl fusion protein

2.4.1 Chronic myeloid leukemia

Deregulation of Abl activity by mutation are usually combined with diverse pathologies such as solid tumors, neurodegenerative diseases, and inflammatory disorders. (Khatri, Wang and Pendergast, 2016) The most common Abl- related diseases is the chronic myelogenous leukemia (CML) , a form of leukemia represented by the unregulated growth of myeloid cells in the bone marrow resulting in sufficient amounts of normal red blood cells, white blood cells, and platelets(Houshmand *et al.*, 2019) (Figure 5).

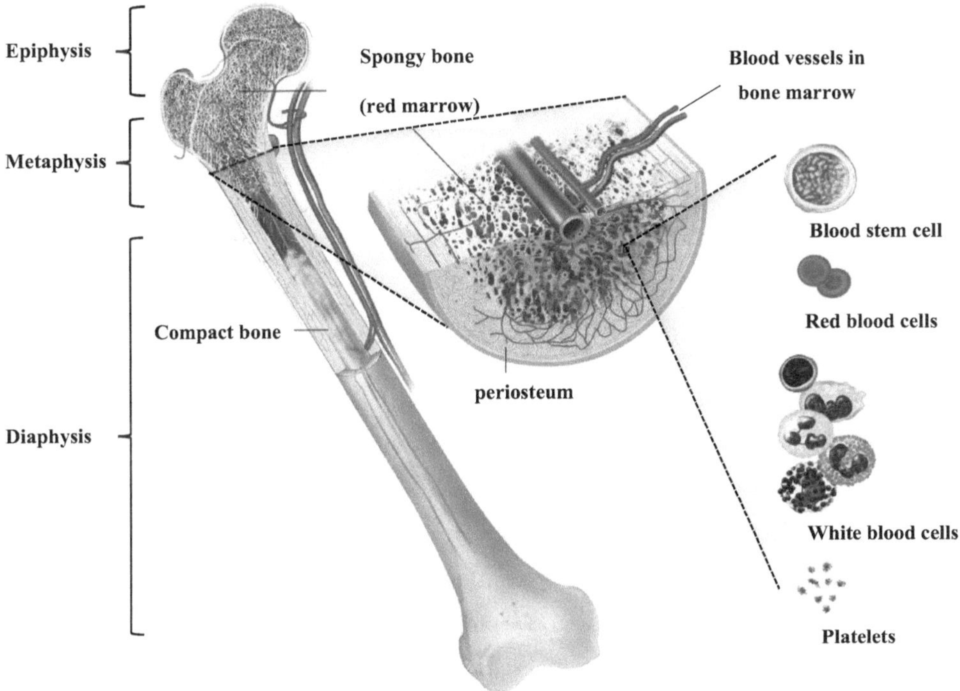

Figure 5: Effect of CML on bone marrow and blood components.

CML is present in about 15-25% of all adult leukemia patients(Granatowicz *et al.*, 2015). The annual incidence rate of CML is 1 to 2 cases per 100 000 and can occur at any age, being most prominent in 60-65 year old (Baccarani *et al.*, 2012) (Figure 6).

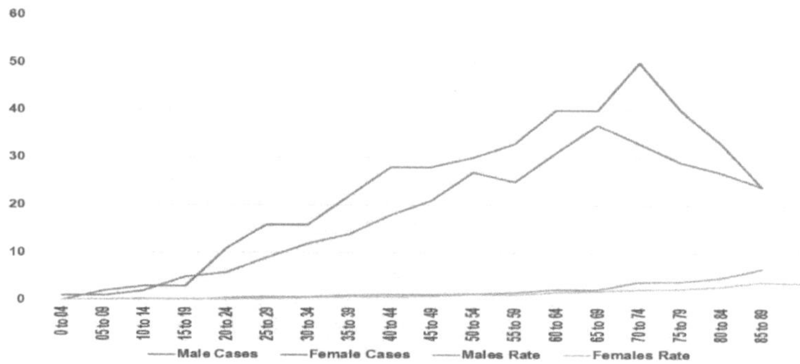

2.4.2 NS3 Protein

The NS3 protein consists of 618 amino acids and a serine-protease domain at its N-terminal and an ATP-driven Helicase domain and RNA triphosphate at its C-terminal. The protein is also involved in viral assembly independently of the known enzymatic activity mentioned (Lescar *et al.*, 2008). During the ZIKV life cycle, the NS3 protein directly interacts with the NS5 polymerase to effectively multiply the viral genome. Studies have shown that impairment of either domain on either protein lead to non-infectious production of the viral particles.

Figure 6: Incidence of Chronic Myeloid Leukemia, Average Number of New Cases per Year and Age-Specific Incidence Rates per 100,000 Population (Prepared by Author).

2.4.2 Bcr-Abl fusion gene:

The expression of the fusion protein (Bcr-Abl) was formed by the conjugation between the Abl1 gene on chromosome 9 with the breakpoint cluster region gene (Bcr) on chromosome 22 generating (Philadelphia chromosome (Ph)) (ROWLEY, 1973). The resulting fusion protein is an integral activated form of the Abl kinase (Bcr-Abl form) that enables unrestrained production of cancerous cells in blood (Wong and Witte, 2004) (Figure 7).

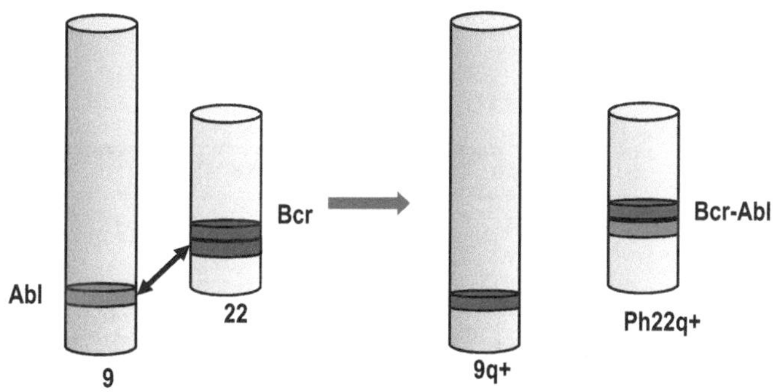

Figure 7: Chromosomal translocation that generates the Philadelphia chromosome and its Bcr-Abl fusion gene (Prepared by Author).

2.4.2 Bcr-Abl: driver of chronic myeloid leukemia:

Bcr-Abl p210 is considered as one of the most common type for Bcr-Abl fusion. it contains all the c-Abl tyrosine kinase sequence except the N-terminal regions which is replaced by Bcr. Absence of N terminal (cap regions) interferes with the myristoylation of the protein and inhabited the release of c-Abl tyrosine kinase from auto inhibition (McWhirter and Wang, 1991) .Abl1 gene can be subdivided into two different splicing transcripts: Abl1a, and Abl1b . The 1b splicing variant is myristoylated at the N-terminus (glycine 2) whereas the 1a variant is 19 amino acids shorter and are not myristoylated (Renshaw, Capozza and Wang, 1988) (Figure 8).

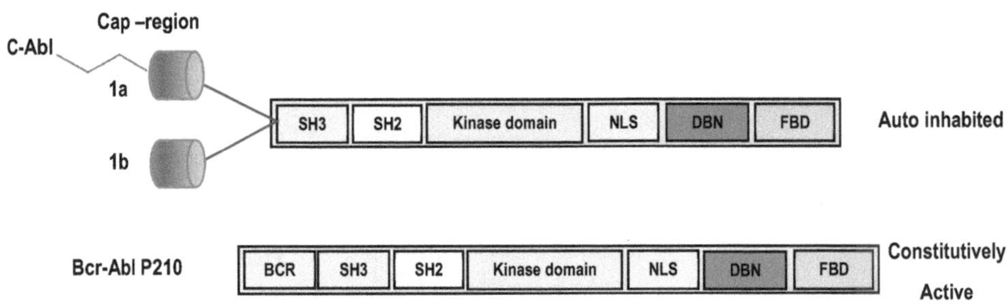

Figure 8: Structural organization of c-Abl and Bcr-Abl. On the top, Domain structures of Abl kinase and on the bottom, the oncogenic Bcr-Abl kinase. **NLS**: Nuclear localization signal, **DBN**: DNA binding domain, **FBD**: F-actin binding domain (Prepared by Author).

c-Abl tyrosine kinase is implicated in diverse range of cellular activities ranging from the regulation of cell growth ,survival ,oxidative stress and

DNA damage response to actin dynamics and cell migration (Van Etten, 1999). The two variants of c-Abl have a modular organization like that Src family members. It is characterized by a core tyrosine kinase domain which is preceded by SH2 and SH3 domains. c-Abl is exclusively having last exon region along carboxy-terminal extension that is containing a nuclear localization signals (NLS) that permit the protein to shuttle between cytoplasm and nucleus (Taagepera *et al.*, 1998) ,and involves in the interaction with F-actin (Hantschel *et al.*, 2005) . It is also having a proline - rich motifs that function as binding sites for SH3 domains of adaptor proteins.(Hantschel *et al.*, 2005) .

2.4.3 Auto-inhibition mechanism:

The tyrosine kinase activity of c-abl is very tightly controlled and c-Abl is mostly inactive in cells (Vihinen *et al.*, 1996; Van Etten, 1999). Structural and biochemical studies have shown a multiple auto inhibitory mechanisms. The auto inhibition relies on a complex set of intramolecular interactions that preserve the kinase in a closed and inactive conformation. The myristoyl group in the cap region has an important role in the autoinhibition of c-Abl 1b (Hantschel *et al.*, 2003). The SH2 domain is formed by a central antiparallel sheet flanked by a α-helices on each side. The mystrate can interact deeply into the hydrophobic pocket in the kinase C-lobe thereforeforcing the SH2 and SH3 domain to dock against the kinase lobes and keeping the kinase inactive (Pluk, Dorey and Superti-Furga, 2002; Nagar *et al.*, 2003) (Figure 9) .

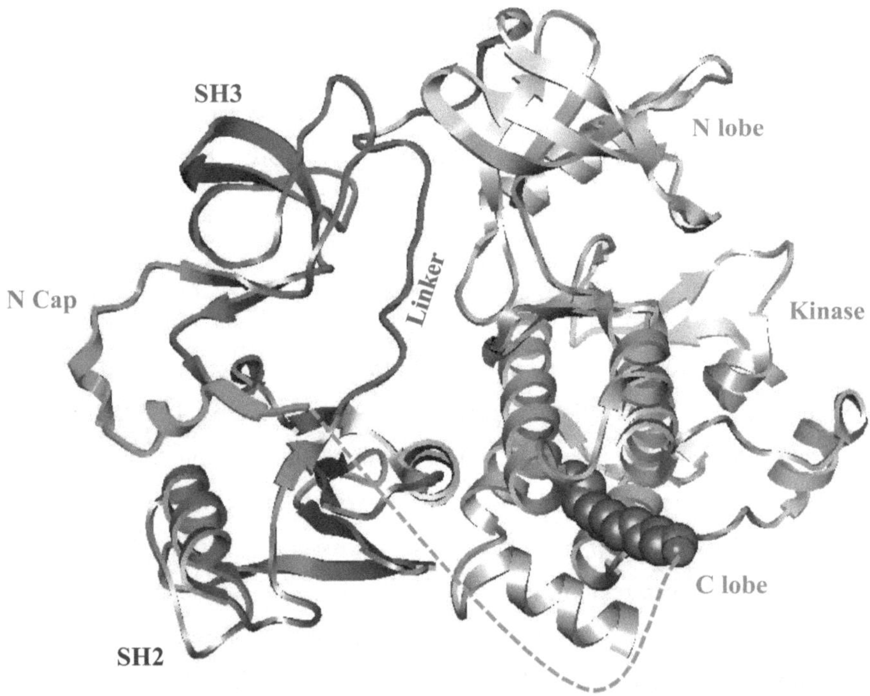

Figure 9 : The 3-D crystal structure of the Bcr-Abl protein (PDB code: 2FO0(Nagar *et al.*, 2006)). The N-cap , SH3, SH2 and kinase domains are

shown in green, orange, blue and gray , respectively. The SH2-kinase linker is depicted in red.

1.5Inhbition of Bcr-Abl

Many of the drugs that are present in market bind to the orthostatic pocket at the ATP binding pocket in Abl kinase (Bcr-Abl). Based on their molecular mechanism of activity , the Bcr-Abl orthostatic inhibitor can be divided into two class, type 1, and type 2 . First generation inhibitors interacts with the active conformation of the kinase, where the catalytic important residues are in the optimal position raised the catalysis, i.e. the DFG in the *in conformation* ,and the Activation -loop is opened. On the other side, the type 2 inhibitors bind to the in active conformation of the Bcr-Abl. i.e the DFG in the *out* conformation, and the Activation -loop is closed or open conformation(Reddy and Aggarwal, 2012) (Figure 10).

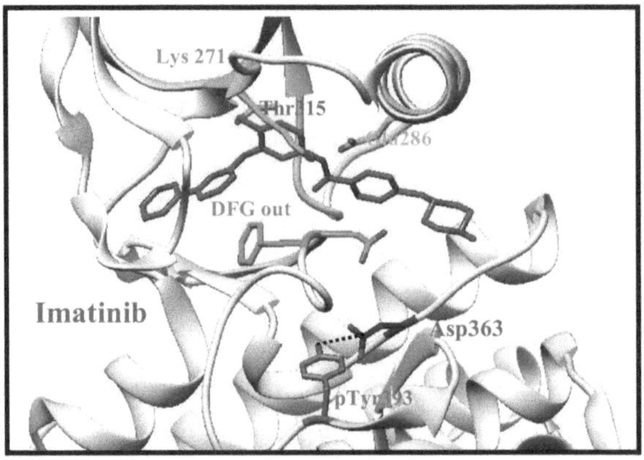

In active Kinase Conformation

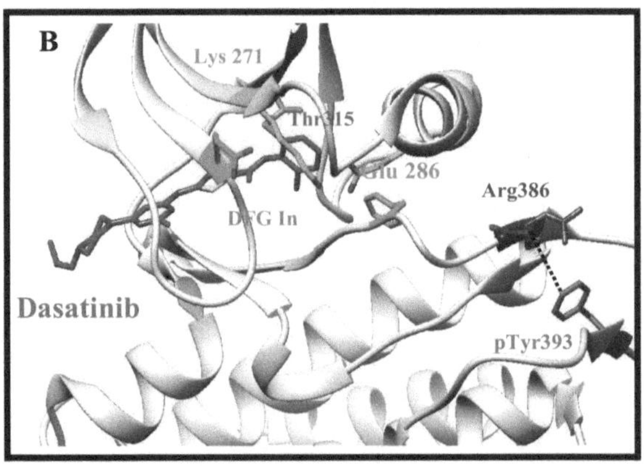

Active kinase conformation

Figure 10: (A) The crystal structure of the Abl kinase domain bound to imatinib (PDB: 1IEP) (Nagar *et al.*, 2002) shows the important features of the inactive configuration of the active site. The ionic interaction between lys 271 and Glu 286 is maintained while the DFG (Asp-Phe-Gly) motif is flipped out. The activation loop is folded towards the active site and Tyr393 forms a hydrogen bond with the preserved Asp363 (catalytic aspartate). (B) Abl kinase domain crystal structure bind to dasatinib (PDB: 2GQG, (Tokarski *et al.*, 2006) , shows significant characteristics of active kinase conformation. To coordinate the ATP phosphate group, the ionic interaction between Lys271 and Glu286 is essential. The phosphorylated Tyr393 forms an Arg386 hydrogen bond and helps stabilize the active conformation.

2.5.1 Targeted therapy and development of resistances

Imatinib (STI-571 ,or CGP-57148) is the first Bcr-Abl inhibitor which was approved in the year 2001 as the drug of choice for the treatment of CML (Zuccotto *et al.*, 2010). Imatinib is a type 2 inhibitors with IC_{50} of 10 nM (Zimmermann *et al.*, 1997) . Nevertheless, the extended mutation across the Abl protein, makes this drug poorly or incompletely active against certain enzymes. The most common mutation are H395P/R ,G250E , M351T,F359V, , M244V, F317L, Y253H/F, Q252H, E355G, E255K/V and gatekeeper T315I. The second generation of Bcr-Abl inhibitors such as Nilotinib and Dasatinib have been approved for the treatment of imatinib – resistant patients with the exception of the gate keeper T315I. (Figure 2.11). Nilotinib (Tasigna, AMN107; Novartis) is a type 2 inhibitors with IC_{50} of 38nM.In 2012, FDA has approved Ponatinib as a third generation ATP competitive tyrosine kinase inhibitor that is efficiently targets for both

Bcr-Abl wild type and the Bcr-Abl T315I mutation with IC_{50}=0.37nM,2.0nM respectively (Figure 2.10) (O'Hare *et al.*, 2009).

Allosteric binding inhibitors have also been found to inhibit the Abl activation, for instance , DCC-2036 is switch pocket inhibitor that allosterically binds to the Arg386/Glu282 residues in the switch regions and hinder Abl from adopting an active conformation(Chan *et al.*, 2011). DCC-2036 has been foundto overcome most of imatinib mutation including T315I mutation. DCC-2036 is now under Phase I clinical trials with T315I mutation CML patients,with the study currently awaiting results (Figure 2.11)(Chan *et al.*, 2011).

other examples include GNF-2 ,and GNF-5 which were able to inhibit Bcr-Abl growth with the exception of T315I mutation (Webersinke, 2016). Similarly, ABL001 is allosterically bound to Bcr-Abl with IC_{50} = 1-12nM (Schoepfer *et al.*, 2018), recent studies have shown that monotherapy using ABL001 which led to tumor regression in mice xenografted with KCL22 CML cell line. Unfortunately the tumors eventually recurred (Figure 2.11) (Wylie *et al.*, 2014, 2017; Eadie *et al.*, 2015)

However, dual targeting of GNF-2 with Nilotinib or ABL001 with Nilotinib have shown a highly synergistic effect compared to mono therapeutic treatment, being able to overcome gatekeeper T315I mutation. Therefore dual targeting of both ATP binding r, and allosteric inhibitors represents an innovative way to overcome mono therapeutic resistance (Iacob *et al.*, 2011) (Wylie *et al.*, 2017). This may be attributed to the effect of GNF5 when bound to the myristate pocket stabilizing the in active form of the T315I mutant, and the new state enable confirmation where I 315 no longer limits access of Nilotinib to the hydrophobic binding pocket . In this case, the allosteric targeting of Abl can cause dynamic perturbations to the ATP binding site residues and this explains the mechanism by which synergistic interaction occurs (Zhang *et al.*, 2010) .

Recently, Mono bodies shows a highly allosterically inhibition to Bcr-Abl by interfering with the SH2-Kinase domain complex. This induces the apoptosis in CML cell lines and primary human CML cells.(Hantschel, Grebien and Superti-Furga, 2012)

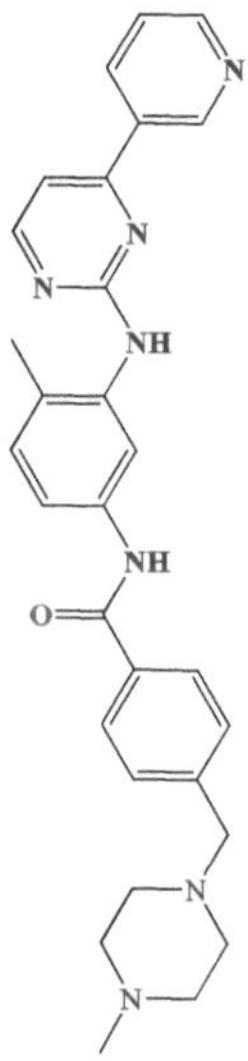

Imatinib, Gleevec, STI-571

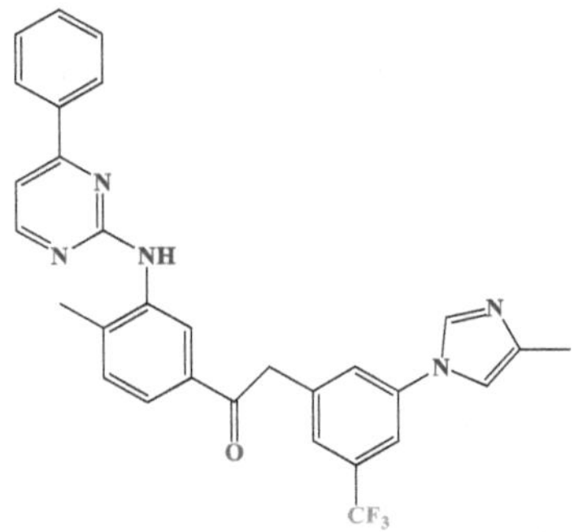

Nilotinib AMN 107

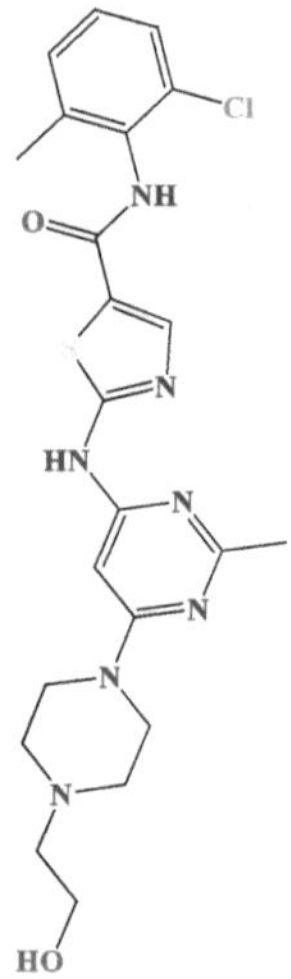

Dasatinib, BMS-354825

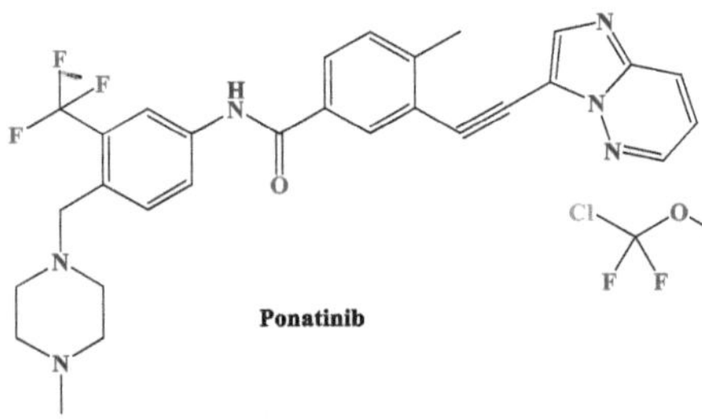

Ponatinib

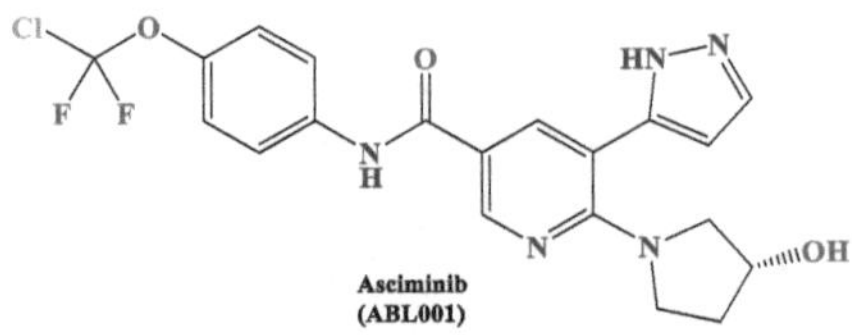

Asciminib (ABL001)

Figure 11: 2D structures of some kinase inhibitors.

1.6Clinical Developments of Protein kinase inhibitors:

Protein kinase inhibitors are rapidly growingin drug discovery, as protein kinases are considered as the second most significant group of drug targets after G-protein-coupled receptors they account for about 20–30% of the drug discovery program of many companies (Vlahovic and Crawford, 2003)(Cohen, 2002a). Recently, FDA has approved 49 compounds which have been identified as protein kinase inhibitors(Table 1) . Despite these encouraging results, response to these protein kinase inhibitors are surrounded by the development of resistance, toxicity ,and efficiency which remains challengeable (Fabbro, Cowan-Jacob and Moebitz, 2015) .Furthermore, Accomplishing the selectively of the kinase inhibitors for particular protein kinases remain a significant challenge(DAVIES et al., 2000; Noble, Endicott and Johnson, 2004). Computational methods can have a key effect on Protein kinase inhibitors design and will stand to be an important feature for the development of these inhibitors (Ferrè, Palmeri and Helmer-Citterich, 2014), so that , many researcher have been motivated to overcome various limitation of kinase inhibition , particularly evading the treatment-related drug resistance .

Table 1: list of FDA-approved small molecule protein kinase inhibitors.

Name	Structure	Target	Domain
Abemacicli b	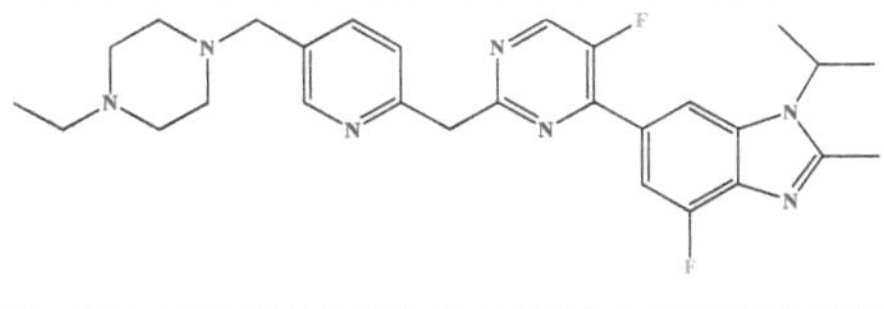	**CDK4/6**	**Breast cancer** (Lu, 2015)
Acalabruti ib	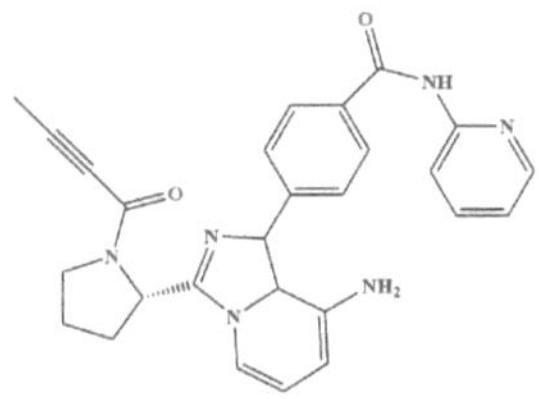	**Bruton tyrosine kinase**	**Mantle cell lymphoma**(Markham and Dhillon, 2018)()
Afatinib	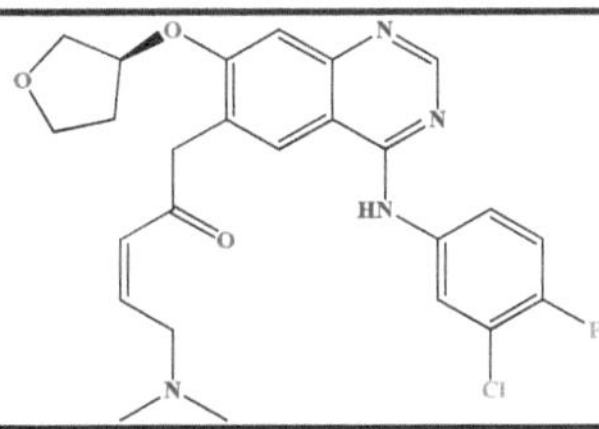	**EGFR, ErbB2, ErbB4**	**NSCLC** (Minkovsky and Berezov, 2008)
Alectinib	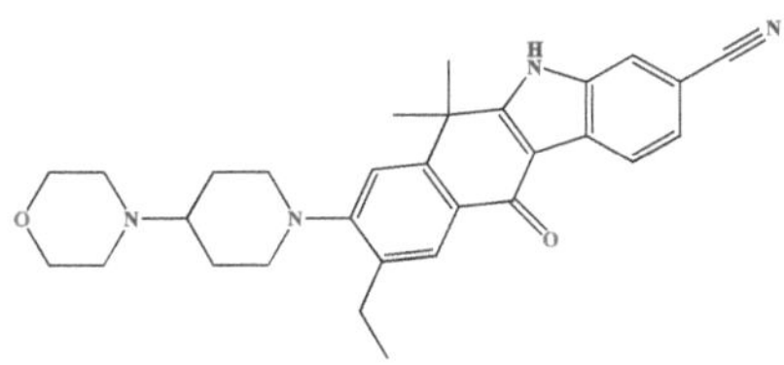	**ALK , RET**	**Anaplastic lymphoma kinase** (McKeage, 2015)
Baricitinib	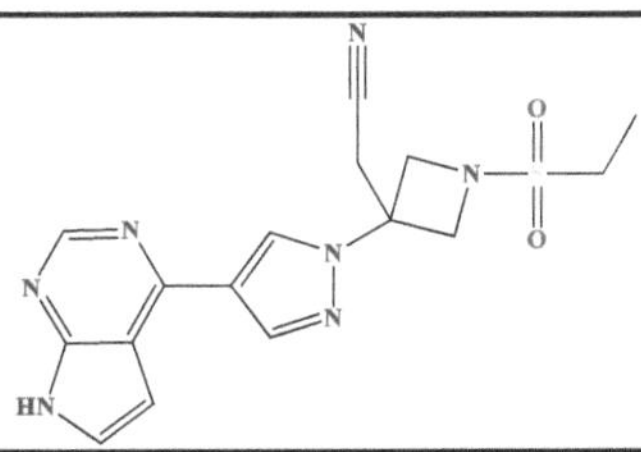	**JAK1/2**	**Rheumatoid arthritis** (Richez *et al.*, 2017)

Binimetinib	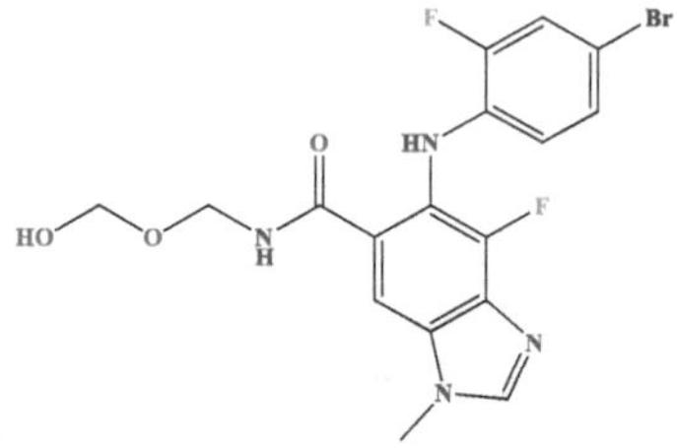	**MEK1/2**	**Cancer** (Koelblinger, Dornbierer and Dummer, 2017)
Bosutinib	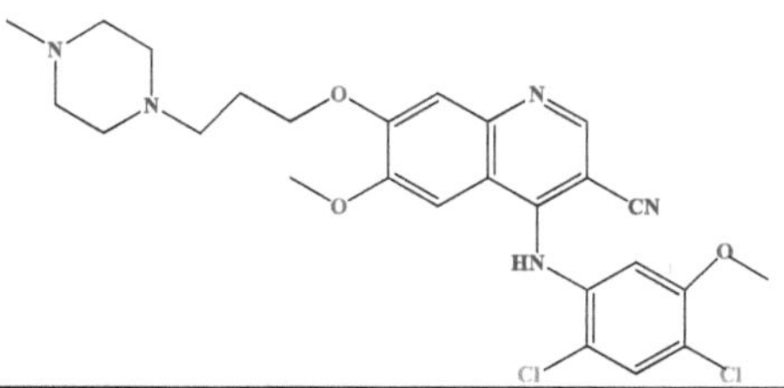	**Bcr-Abl, Src, Lyn, Hck**	**CML**(Daud *et al.*, 2012)
Brigatinib	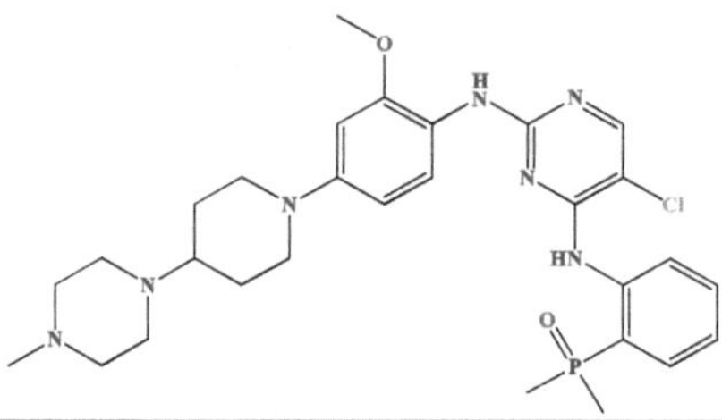	**ALK, EGFR**	**Cancer** (Huang *et al.*, 2016)
Cobozantinib	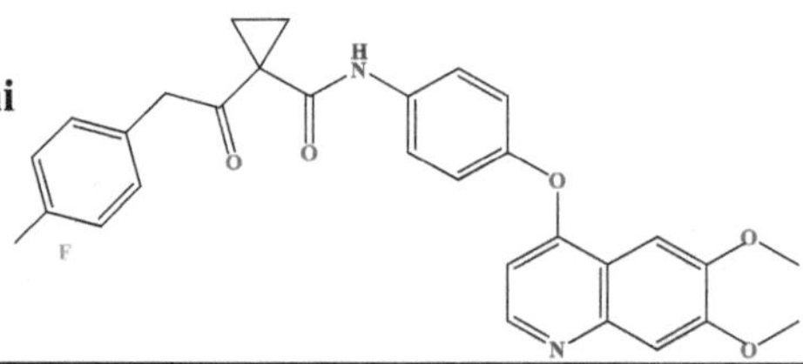	**Met, VEGFR2, Axl**	**Cancer** (Choueiri *et al.*, 2015)

Ceritinib	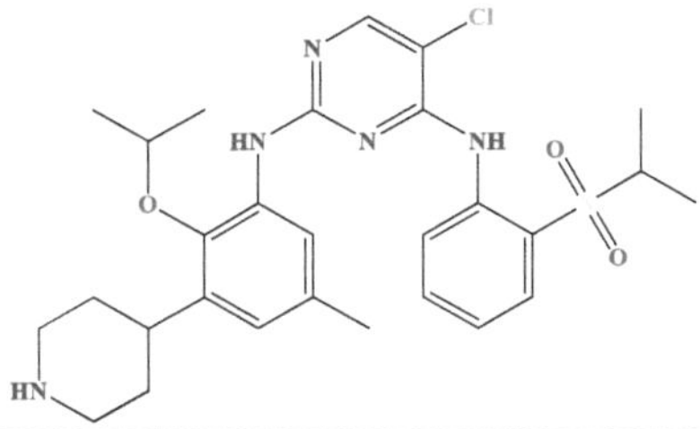	**ALK**	**NSCLC** (Soria *et al.*, 2017)
Dacomitini	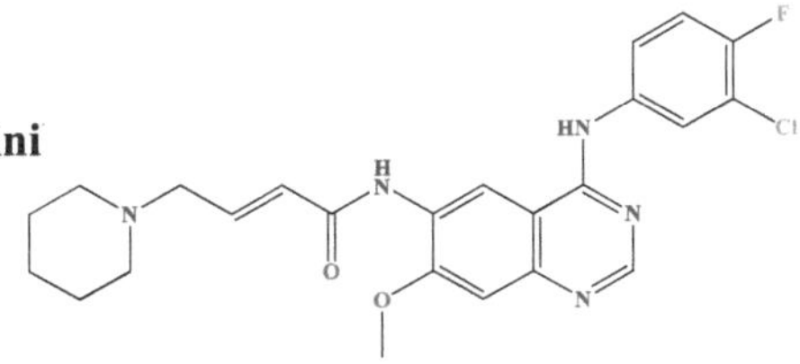	**EGFR**	**NSCLC**(Shirley, 2018)
Dasatinib	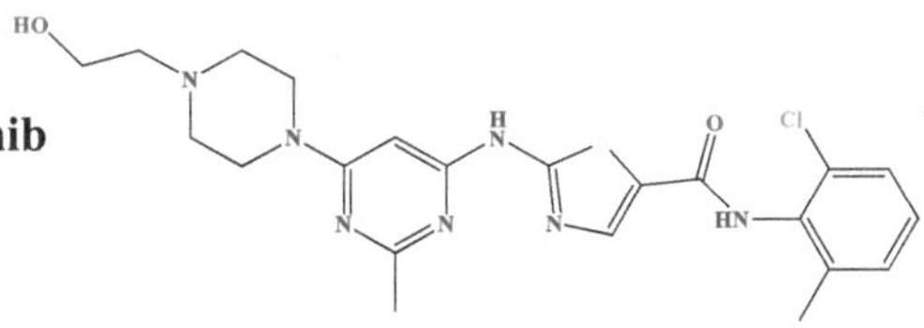	**Bcr-Abl**	**CML**(Tokarski *et al.*, 2006)
Encorafeni	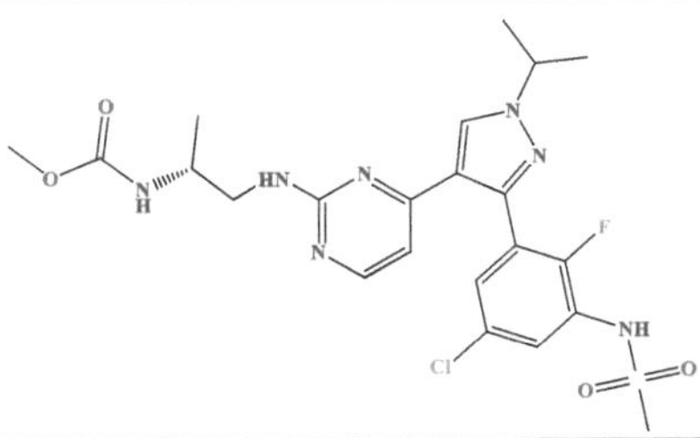	**B-Raf$^{V600E/K}$**	**Cancer** (Koelblinger, Thuerigen and Dummer, 2018)

Fostamatinib	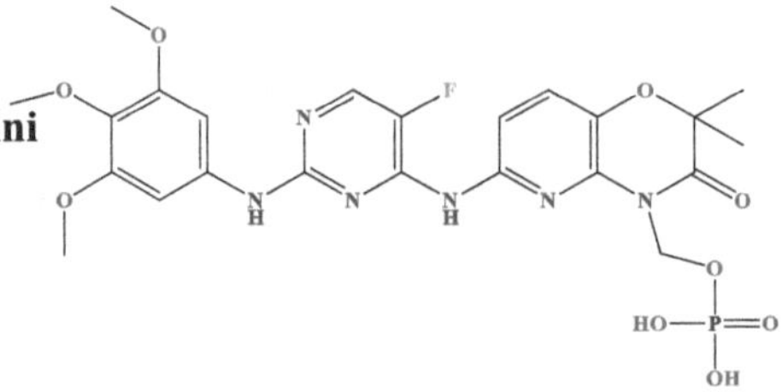	**Spleen tyrosine kinase (syk)**	**Autoimmune disorders**(Braselmann *et al.*, 2006)
Gilteritinib	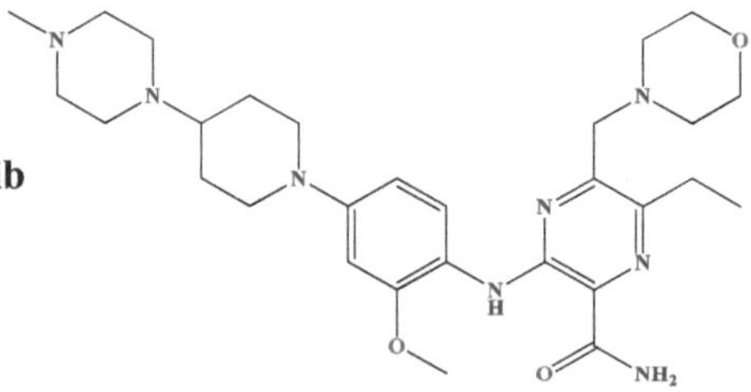	**FLT3 Mutation**	**AML** (Perl *et al.*, 2017)
Larotrectinib	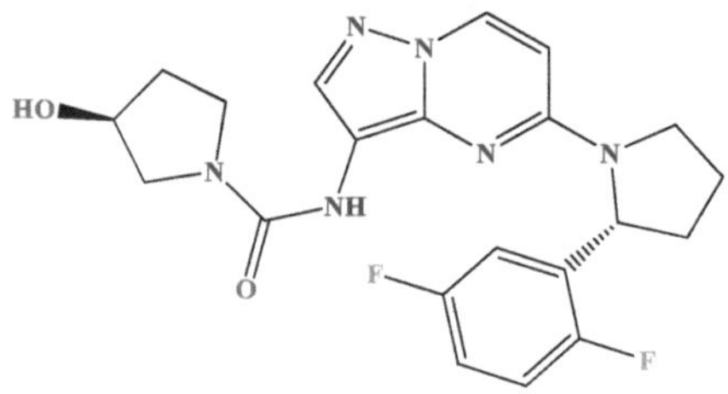	**NTRK**	**Cancer** (Berger, Martens and Bochum, 2018)
Lorlatinib	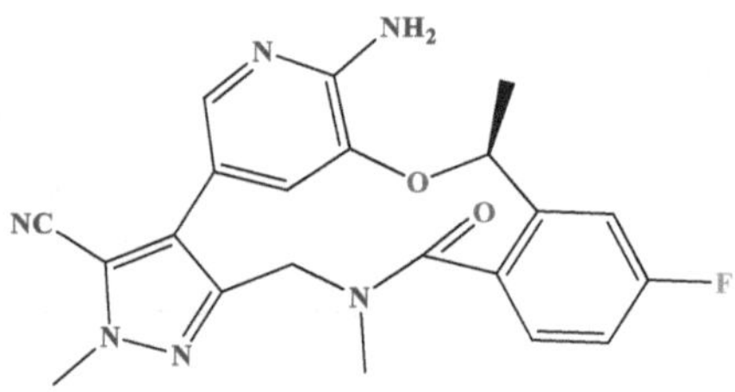	**ALK**	**NSCLC** (Shaw *et al.*, 2017)

Netarsudil	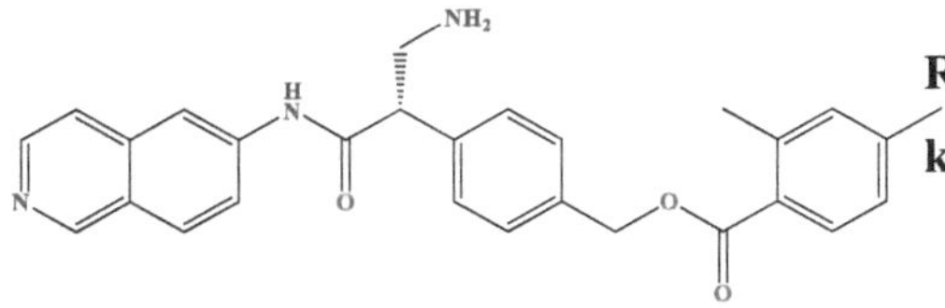	**Rho kinase**	**Ocular hypertension** (Lin *et al.*, 2018)
Imatinib	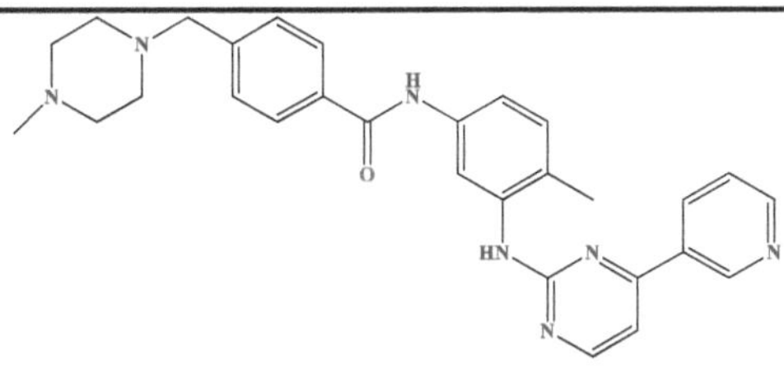	**Bcr-Abl kinase**	CML (Capdeville *et al.*, 2002)
Lenvatinib	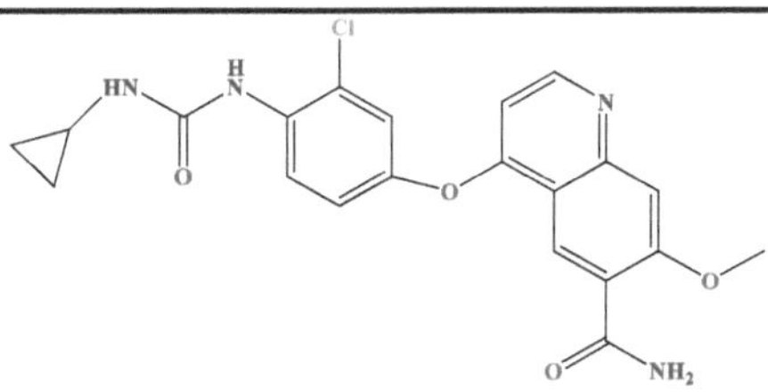	**VEGFRs, FGFRs, PDGFR, Kit, RET**	**Cancer** (Matsui *et al.*, 2008)
Ponatinib	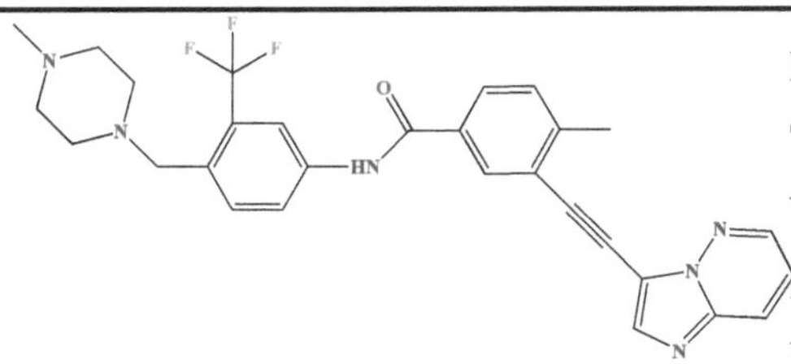	**Bcr-Abl, T315I, VEGFR, PDGFR, FGFR**	**Cancer** (Huang *et al.*, 2010)

Nilotinib	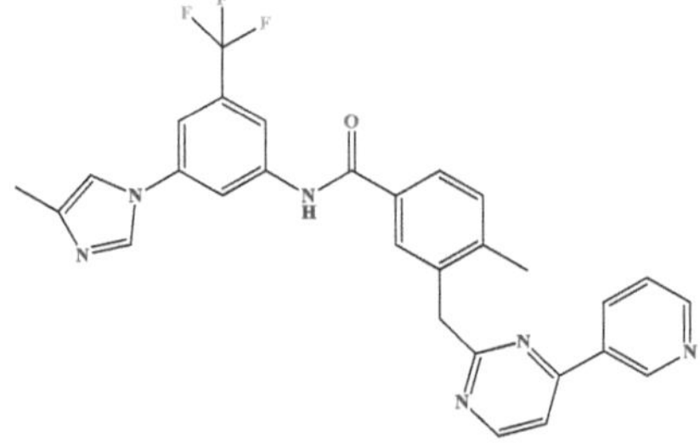	**BCR-Abl, PDGFR,**	**CML** (Golemovic *et al.*, 2005)
Neratinib	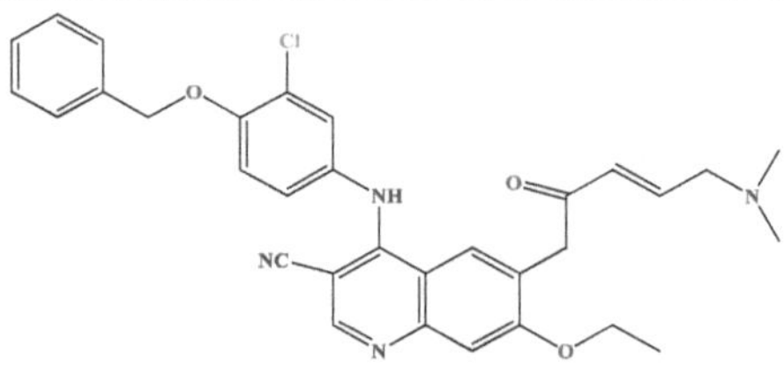	**ErbB2/HER2**	**Cancer** (Gandhi *et al.*, 2017)
Osimertinib	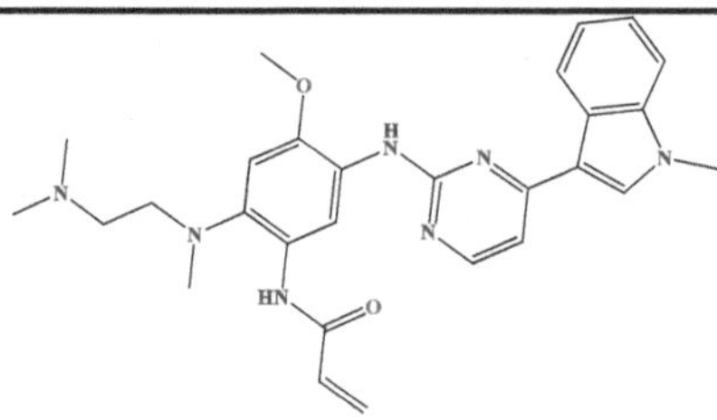	**EGFR T970M**	**NSCLC** (Ayeni, Politi and Goldberg, 2015)
Palbociclib	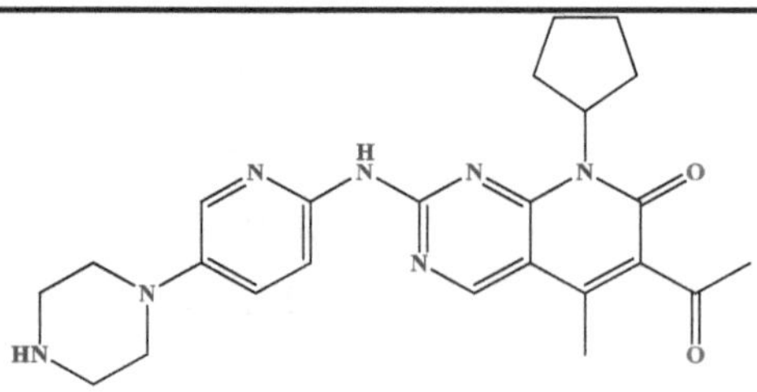	**CDK4/6**	**Cancer** (Finn *et al.*, 2009)

Pazopanib	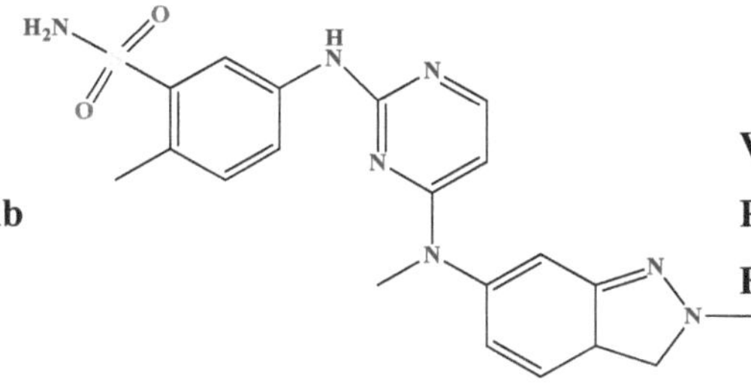	**VEGFR, PDGFR, FGFR**	**Cancer** (Verheijen *et al.*, 2017)
Ribociclib	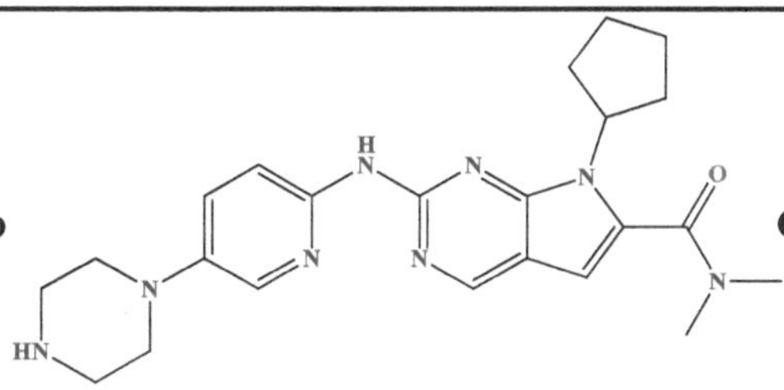	**CDK4/6**	**Cancer**(Samson, 2014)

Conclusion :

The journey of targeting the Bcr-Abl kinase in Chronic Myeloid Leukemia stands as a transformative paradigm in oncology, illustrating the powerful convergence of basic molecular biology and rational drug design. From the initial discovery of the Philadelphia chromosome to the clinical deployment of imatinib, this story marked a shift from non-specific cytotoxic chemotherapy to precision medicine, offering patients unprecedented survival and quality of life.

However, the adaptability of cancer, manifested through point mutations like T315I, revealed a critical limitation of ATP-competitive inhibitors and underscored the perpetual challenge of drug resistance. This challenge has, in turn, fueled a new wave of innovation, driving the development of more potent ATP-competitive agents and, most significantly, the pioneering class of allosteric inhibitors. Compounds such as asciminib (ABL001), which target the myristoyl pocket, represent a strategic evolution in our approach. By exploiting a distinct regulatory site, they not only circumvent common resistance mutations but also open the door for synergistic combination therapies. The strategy of dual targeting—simultaneously inhibiting the ATP-binding site and an allosteric site—has proven highly effective in overcoming resistance and presents a robust blueprint for suppressing the emergence of resistant clones.

The lessons learned from the Bcr-Abl saga extend far beyond CML. They provide a master template for targeting other oncogenic kinases and underscore fundamental principles in drug discovery: the critical importance of understanding protein structure and autoinhibitory mechanisms, the need to anticipate and preempt resistance mechanisms, and the immense therapeutic potential of moving beyond the ATP-binding cleft. As the field advances, the integration of computational biology, structural insights, and combinatorial strategies will be crucial in developing the next generation of kinase inhibitors. The success achieved in CML is not an endpoint but a beacon, guiding the continued pursuit of targeted, intelligent, and durable therapies for a broader spectrum of cancers.

Reference

Abou-Jawde, R. *et al.* (2003) ‗An overview of targeted treatments in cancer', *Clinical Therapeutics*. Elsevier, 25(8), pp. 2121–2137. doi: 10.1016/S0149-2918(03)80209-6.

Anand, P. *et al.* (2008) ‗Cancer is a Preventable Disease that Requires Major Lifestyle Changes', *Pharmaceutical Research*, 25(9), pp. 2097–2116. doi: 10.1007/s11095-008-9661-9.

Ayeni, D., Politi, K. and Goldberg, S. B. (2015) ‗Emerging Agents and New Mutations in EGFR-Mutant Lung Cancer', *Clinical Cancer Research*, 21(17), pp. 3818–3820. doi: 10.1158/1078-0432.CCR-15-1211.

Baccarani, M. *et al.* (2012) ‗Chronic myeloid leukemia: ESMO Clinical Practice Guidelines for diagnosis, treatment and follow-up', *Annals of Oncology*. Karger, Basel, 23(suppl 7), pp. 72–77. doi: 10.1093/annonc/mds228.

Berger, S., Martens, U. M. and Bochum, S. (2018) ‗Larotrectinib (LOXO-101)', in *Recent results in cancer research. Fortschritte der Krebsforschung. Progres dans les recherches sur le cancer*, pp. 141–151. doi: 10.1007/978-3-319-91442-8_10.

Blume-Jensen, P. and Hunter, T. (2001) ‗Oncogenic kinase signalling.', *Nature*, 411(6835), pp. 355–365. doi: 10.1038/35077225.

Braselmann, S. *et al.* (2006) ‗R406, an Orally Available Spleen Tyrosine Kinase Inhibitor Blocks Fc Receptor Signaling and Reduces Immune Complex-Mediated Inflammation', *Journal of Pharmacology and Experimental Therapeutics*, 319(3), pp. 998–1008. doi:

10.1124/jpet.106.109058.

Capdeville, R. *et al.* (2002) ‗Glivec (STI571, imatinib), a rationally developed, targeted anticancer drug.', *Nature reviews. Drug discovery*, 1(7), pp. 493–502. doi: 10.1038/nrd839.

Chan, W. W. *et al.* (2011) ‗Conformational control inhibition of the BCR-ABL1 tyrosine kinase, including the gatekeeper T315I mutant, by the switch-control inhibitor DCC-2036.', *Cancer cell*, 19(4), pp. 556–68. doi: 10.1016/j.ccr.2011.03.003.

Choueiri, T. K. *et al.* (2015) ‗Cabozantinib versus everolimus in advanced renal-cell carcinoma', *New England Journal of Medicine*. Mass Medical Soc, 373(19), pp. 1814–1823.

Cohen, P. (2002a) ‗Protein kinases — the major drug targets of the twenty-first century?', *Nature Reviews Drug Discovery*. Nature Publishing Group, 1(4), pp. 309–315. doi: 10.1038/nrd773.

Cohen, P. (2002b) ‗The origins of protein phosphorylation.', *Nature cell biology*, 4(5), pp. E127-30. doi: 10.1038/ncb0502-e127.

Cox, S., Radzio-Andzelm, E. and Taylor, S. S. (1994) ‗Domain movements in protein kinases', *Current Opinion in Structural Biology*. Elsevier Current Trends, 4(6), pp. 893–901. doi: 10.1016/0959-440X(94)90272-0.

Dar, A. C. and Shokat, K. M. (2011) ‗The Evolution of Protein Kinase Inhibitors from Antagonists to Agonists of Cellular Signaling', *Annual Review of Biochemistry*, 80(1), pp. 769–795. doi: 10.1146/annurev-biochem-090308-173656.

Daud, A. I. *et al.* (2012) ‗Phase I study of bosutinib, a src/abl tyrosine kinase inhibitor, administered to patients with advanced solid tumors‘, *Clinical Cancer Research*. AACR, 18(4), pp. 1092–1100.

DAVIES, S. P. *et al.* (2000) _Specificity and mechanism of action of some commonly used protein kinase inhibitors‘, *Biochemical Journal*, 351(1), p. 95. doi: 10.1042/0264-6021:3510095.

Dhillon, A. S. *et al.* (2007) _MAP kinase signalling pathways in cancer.‘, *Oncogene*, 26(22), pp. 3279–90. doi: 10.1038/sj.onc.1210421.

Eadie, L. N. *et al.* (2015) _The Allosteric Inhibitor ABL001 Is Susceptible to Resistance in Vitro Mediated By Overexpression of the Drug Efflux Transporters ABCB1 and ABCG2‘, *Blood*, 126(23), pp. 4841–4841. Available at: http://www.bloodjournal.org/content/126/23/4841?sso-checked=true (Accessed: 23 February 2018).

Van Etten, R. A. (1999) _Cycling, stressed-out and nervous: cellular functions of c-Abl.‘, *Trends in cell biology*, 9(5), pp. 179–86. Available at: http://www.ncbi.nlm.nih.gov/pubmed/10322452 (Accessed: 4 April 2018).

Fabbro, D., Cowan-Jacob, S. W. and Moebitz, H. (2015) _Ten things you should know about protein kinases: IUPHAR Review 14‘, *British Journal of Pharmacology*, 172(11), pp. 2675–2700. doi: 10.1111/bph.13096.

Ferrè, F., Palmeri, A. and Helmer-Citterich, M. (2014) _Computational methods for analysis and inference of kinase/inhibitor relationships.‘, *Frontiers in genetics*. Frontiers Media SA, 5, p. 196. doi: 10.3389/fgene.2014.00196.

Finn, R. S. *et al.* (2009) _PD 0332991, a selective cyclin D kinase 4/6 inhibitor, preferentially inhibits proliferation of luminal estrogen receptor-positive human breast cancer cell lines in vitro‘, *Breast Cancer Research*, 11(5), p. R77. doi: 10.1186/bcr2419.

Gandhi, L. *et al.* (2017) ‚MA04.02 Neratinib ± Temsirolimus in HER2-Mutant Lung Cancers: An International, Randomized Phase II Study‘, *Journal of Thoracic Oncology*. Elsevier, 12(1), pp. S358–S359. doi: 10.1016/J.JTHO.2016.11.398.

Golemovic, M. *et al.* (2005) ‚AMN107, a novel aminopyrimidine inhibitor of Bcr-Abl, has in vitro activity against imatinib-resistant chronic myeloid leukemia.‘, *Clinical cancer research : an official journal of the American Association for Cancer Research*. American Association for Cancer Research, 11(13), pp. 4941–7. doi: 10.1158/1078-0432.CCR-04-2601.

Granatowicz, A. *et al.* (2015) ‚An Overview and Update of Chronic Myeloid Leukemia for Primary Care Physicians.‘, *Korean journal of family medicine*. Korean Academy of Family Medicine, 36(5), pp. 197–202. doi: 10.4082/kjfm.2015.36.5.197.

Hantschel, O. *et al.* (2003) ‚A myristoyl/phosphotyrosine switch regulates c-Abl.‘, *Cell*, 112(6), pp. 845–57. Available at: http://www.ncbi.nlm.nih.gov/pubmed/12654250 (Accessed: 12 February 2018).

Hantschel, O. *et al.* (2005) ‚Structural Basis for the Cytoskeletal Association of Bcr-Abl/c-Abl‘, *Molecular Cell*, 19(4), pp. 461–473. doi: 10.1016/j.molcel.2005.06.030.

Hantschel, O., Grebien, F. and Superti-Furga, G. (2012) ‚The growing arsenal of ATP-competitive and allosteric inhibitors of BCR-ABL.‘, *Cancer research*, 72(19), pp. 4890–5. doi: 10.1158/0008-5472.CAN-12-1276.

Houshmand, M. *et al.* (2019) ‗Chronic myeloid leukemia stem cells', *Leukemia*. Nature Publishing Group, 33(7), pp. 1543–1556. doi: 10.1038/s41375-019-0490-0.

Huang, W.-S. *et al.* (2010) ‗Discovery of 3-[2-(imidazo[1,2-b]pyridazin-3-yl)ethynyl]-4-methyl-N-{4-[(4-methylpiperazin-1-yl)methyl]-3-(trifluoromethyl)phenyl}benzamide (AP24534), a potent, orally active pan-inhibitor of breakpoint cluster region-abelson (BCR-ABL) kinase including the T315I gatekeeper mutant.', *Journal of medicinal chemistry*, 53(12), pp. 4701–19. doi: 10.1021/jm100395q.

Huang, W.-S. *et al.* (2016) ‗Discovery of brigatinib (AP26113), a phosphine oxide-containing, potent, orally active inhibitor of anaplastic lymphoma kinase', *Journal of medicinal chemistry*. ACS Publications, 59(10), pp. 4948–4964.

Hubbard, S. R., and Till, J. H. (2000) ‗Protein Tyrosine Kinase Structure and Function', *Annual Review of Biochemistry*, 69(1), pp. 373–398. doi: 10.1146/annurev.biochem.69.1.373.

Hunter, T. (2012) ‗Why nature chose phosphate to modify proteins.', *Philosophical transactions of the Royal Society of London. Series B, Biological sciences*. The Royal Society, 367(1602), pp. 2513–6. doi: 10.1098/rstb.2012.0013.

Huse, M. and Kuriyan, J. (2002) ‗The Conformational Plasticity of Protein Kinases', *Cell*. Cell Press, 109(3), pp. 275–282. doi: 10.1016/S0092-8674(02)00741-9.

Iacob, R. E. *et al.* (2011) ‗Allosteric interactions between the myristate- and ATP-site of the Abl kinase.', *PloS one*. Edited by F. Rodrigues-Lima, 6(1),

p. e15929. doi: 10.1371/journal.pone.0015929.

Johnson, L. N. and Lewis, R. J. (2001) ‗Structural Basis for Control by Phosphorylation', *Chemical Reviews*, 101(8), pp. 2209–2242. doi: 10.1021/cr000225s.

Kang, J.-H. (2014) ‗Protein Kinase C (PKC) Isozymes and Cancer', *New Journal of Science*. Hindawi, 2014, pp. 1–36. doi: 10.1155/2014/231418.

Kato-Stankiewicz, J. *et al.* (2001) ‗Epidermal Growth Factor Stimulation of the ACK1/Dbl Pathway in a Cdc42 and Grb2-Dependent Manner', *Biochemical and Biophysical Research Communications*. Academic Press, 284(2), pp. 470–477. doi: 10.1006/BBRC.2001.5004.

Khatri, A., Wang, J. and Pendergast, A. M. (2016) ‗Multifunctional Abl kinases in health and disease.', *Journal of cell science*, 129(1), pp. 9–16. doi: 10.1242/jcs.175521.

Kisqali - FDA prescribing information, side effects and uses (no date). Available at: https://www.drugs.com/pro/kisqali.html (Accessed: 1 August 2019).

Koelblinger, P., Dornbierer, J. and Dummer, R. (2017) ‗A review of binimetinib for the treatment of mutant cutaneous melanoma', *Future Oncology*, 13(20), pp. 1755–1766. doi: 10.2217/fon-2017-0170.

Koelblinger, P., Thuerigen, O. and Dummer, R. (2018) ‗Development of encorafenib for BRAF-mutated advanced melanoma', *Current Opinion in Oncology*, 30(2), p. 1. doi: 10.1097/CCO.0000000000000426.

Kornev, A. P. *et al.* (2006) ‗Surface comparison of active and inactive protein kinases identifies a conserved activation mechanism.‘, *Proceedings of the National Academy of Sciences of the United States of America*. National Academy of Sciences, 103(47), pp. 17783–8. doi: 10.1073/pnas.0607656103.

Kornev, A. P. and Taylor, S. S. (2015) ‗Dynamics-Driven Allostery in Protein Kinases.‘, *Trends in biochemical sciences*, 40(11), pp. 628–647. doi: 10.1016/j.tibs.2015.09.002.

Kung, J. E. and Jura, N. (2016) ‗Structural Basis for the Non-catalytic Functions of Protein Kinases.‘, *Structure (London, England: 1993)*. Elsevier, 24(1), pp. 7–24. doi: 10.1016/j.str.2015.10.020.

Kuriyan, J. and Eisenberg, D. (2007) ‗The origin of protein interactions and allostery in colocalization.‘, *Nature*, 450(7172), pp. 983–90. doi: 10.1038/nature06524.

Leonard, T. A. and Hurley, J. H. (2011) ‗Regulation of protein kinases by lipids.‘, *Current opinion in structural biology*. NIH Public Access, 21(6), pp. 785–91. doi: 10.1016/j.sbi.2011.07.006.

Lin, C.-W. *et al.* (2018) ‗Discovery and Preclinical Development of Netarsudil, a Novel Ocular Hypotensive Agent for the Treatment of Glaucoma‘, *Journal of Ocular Pharmacology and Therapeutics*, 34(1–2), pp. 40–51. doi: 10.1089/jop.2017.0023.

Lu, J. (2015) ‗Palbociclib: a first-in-class CDK4/CDK6 inhibitor for the treatment of hormone-receptor positive advanced breast cancer.‘, *Journal of hematology & oncology*. BioMed Central, 8, p. 98. doi: 10.1186/s13045-015-0194-5.

Manning, G. *et al.* (2002) ‗The Protein Kinase Complement of the Human Genome', *Science*, 298(5600), pp. 1912–1934. doi: 10.1126/science.1075762.

Markham, A. and Dhillon, S. (2018) ‗Acalabrutinib: First Global Approval', *Drugs*, 78(1), pp. 139–145. doi: 10.1007/s40265-017-0852-8.

Masterson, L. R. *et al.* (2012) ‗Allostery and Binding Cooperativity of the Catalytic Subunit of Protein Kinase A by NMR Spectroscopy and Molecular Dynamics Simulations', *Advances in Protein Chemistry and Structural Biology*. Academic Press, 87, pp. 363–389. doi: 10.1016/B978-0-12-398312-1.00012-3.

Matsui, J. *et al.* (2008) ‗Multi-kinase inhibitor E7080 suppresses lymph node and lung metastases of human mammary breast tumor MDA-MB-231 via inhibition of vascular endothelial growth factor-receptor (VEGF-R) 2 and VEGF-R3 kinase.', *Clinical cancer research : an official journal of the American Association for Cancer Research*, 14(17), pp. 5459–65. doi: 10.1158/1078-0432.CCR-07-5270.

McKeage, K. (2015) ‗Alectinib: A Review of Its Use in Advanced ALK-Rearranged Non-Small Cell Lung Cancer', *Drugs*, 75(1), pp. 75–82. doi: 10.1007/s40265-014-0329-y.

McWhirter, J. R. and Wang, J. Y. (1991) ‗Activation of tyrosinase kinase and microfilament-binding functions of c-abl by bcr sequences in bcr/abl fusion proteins.', *Molecular and cellular biology*. American Society for Microbiology (ASM), 11(3), pp. 1553–65. Available at: http://www.ncbi.nlm.nih.gov/pubmed/1705008 (Accessed: 5 April 2018).

Minkovsky, N. and Berezov, A. (2008) ‗BIBW-2992, a dual receptor tyrosine kinase inhibitor for the treatment of solid tumors.', *Current opinion in investigational drugs (London, England : 2000)*, 9(12), pp. 1336–46. Available at: http://www.ncbi.nlm.nih.gov/pubmed/19037840 (Accessed: 1 August 2019).

Nagar, B. *et al.* (2002) _Crystal structures of the kinase domain of c-Abl in complex with the small molecule inhibitors PD173955 and imatinib (STI-571).', *Cancer research*, 62(15), pp. 4236–43. Available at: http://www.ncbi.nlm.nih.gov/pubmed/12154025 (Accessed: 22 March 2018).

Nagar, B. *et al.* (2003) _Structural basis for the autoinhibition of c-Abl tyrosine kinase.', *Cell*, 112(6), pp. 859–71. Available at: http://www.ncbi.nlm.nih.gov/pubmed/12654251 (Accessed: 18 January 2018).

Nagar, B. *et al.* (2006) _Organization of the SH3-SH2 unit in active and inactive forms of the c-Abl tyrosine kinase.', *Molecular cell*, 21(6), pp. 787–98. doi: 10.1016/j.molcel.2006.01.035.

Neet, K. and Hunter, T. (1996) _Vertebrate non-receptor protein-tyrosine kinase families.', *Genes to cells : devoted to molecular & cellular mechanisms*, 1(2), pp. 147–69. Available at: http://www.ncbi.nlm.nih.gov/pubmed/9140060 (Accessed: 3 April 2018).

Noble, M. E. M., Endicott, J. A. and Johnson, L. N. (2004) _Protein kinase inhibitors: insights into drug design from structure.', *Science (New York, N.Y.)*, 303(5665), pp. 1800–5. doi: 10.1126/science.1095920.

O'Hare, T. *et al.* (2009) _AP24534, a Pan-BCR-ABL Inhibitor for Chronic Myeloid Leukemia, Potently Inhibits the T315I Mutant and Overcomes

Mutation-Based Resistance‘, *Cancer Cell*, 16(5), pp. 401–412. doi: 10.1016/j.ccr.2009.09.028.

Pellicena, P. and Miller, W. T. (2001) ‗Processive Phosphorylation of p130Cas by Src Depends on SH3-Polyproline Interactions‘, *Journal of Biological Chemistry*, 276(30), pp. 28190–28196. doi: 10.1074/jbc.M100055200.

Perl, A. E. *et al.* (2017) ‗Selective inhibition of FLT3 by gilteritinib in relapsed or refractory acute myeloid leukaemia: a multicentre, first-in-human, open-label, phase 1–2 study‘, *The Lancet Oncology*, 18(8), pp. 1061–1075. doi: 10.1016/S1470-2045(17)30416-3.

Pluk, H., Dorey, K. and Superti-Furga, G. (2002) ‗Autoinhibition of c-Abl.‘, *Cell*, 108(2), pp. 247–59. Available at: http://www.ncbi.nlm.nih.gov/pubmed/11832214 (Accessed: 28 June 2017).

Reddy, E. P. and Aggarwal, A. K. (2012) ‗The ins and outs of bcr-abl inhibition.‘, *Genes & cancer*, 3(5–6), pp. 447–54. doi: 10.1177/1947601912462126.

Renshaw, M. W., Capozza, M. A. and Wang, J. Y. (1988) ‗Differential expression of type-specific c-abl mRNAs in mouse tissues and cell lines.‘, *Molecular and cellular biology*, 8(10), pp. 4547–51. Available at: http://www.ncbi.nlm.nih.gov/pubmed/2460747 (Accessed: 4 April 2018).

Richez, C. *et al.* (2017) ‗Efficacy of baricitinib in the treatment of rheumatoid arthritis‘, *Expert Opinion on Pharmacotherapy*, 18(13), pp. 1399–1407. doi: 10.1080/14656566.2017.1359256.

Robinson, D. R., Wu, Y.-M. and Lin, S.-F. (2000) ‗The protein tyrosine kinase family of the human genome‘, *Oncogene*, 19(49), pp. 5548–5557. doi: 10.1038/sj.onc.1203957.

ROWLEY, J. D. (1973) ‗A New Consistent Chromosomal Abnormality in Chronic Myelogenous Leukaemia identified by Quinacrine Fluorescence and Giemsa Staining‘, *Nature*. Nature Publishing Group, 243(5405), pp. 290–293. doi: 10.1038/243290a0.

S, M. and S, M. (2015) *Novel Approaches and Strategies for Biologics, Vaccines and Cancer Therapies*. Elsevier. doi: 10.1016/C2013-0-00324-6.

Samson, K. (2014) ‗LEE011 CDK Inhibitor Showing Early Promise in Drug-Resistant Cancers‘, *Oncology Times*, 36(3), pp. 39–40. doi: 10.1097/01.COT.0000444043.33304.c1.

Schoepfer, J. *et al.* (2018) ‗Discovery of Asciminib (ABL001), an Allosteric Inhibitor of the Tyrosine Kinase Activity of BCR-ABL1‘, *Journal of Medicinal Chemistry*. American Chemical Society, 61(18), pp. 8120–8135. doi: 10.1021/acs.jmedchem.8b01040.

Shaw, A. T. *et al.* (2017) ‗Lorlatinib in non-small-cell lung cancer with ALK or ROS1 rearrangement: an international, multicentre, open-label, single-arm first-in-man phase 1 trial‘, *The Lancet Oncology*, 18(12), pp. 1590–1599. doi: 10.1016/S1470-2045(17)30680-0.

Shirley, M. (2018) ‗Dacomitinib: First Global Approval‘, *Drugs*, 78(18), pp. 1947–1953. doi: 10.1007/s40265-018-1028-x.

Sirvent, A., Benistant, C. and Roche, S. (2008) ‗Cytoplasmic signalling by the c-Abl tyrosine kinase in normal and cancer cells‘, *Biol. Cell*, 100, pp. 617–631. doi: 10.1042/BC20080020.

Soria, J.-C. *et al.* (2017) ‗First-line ceritinib versus platinum-based chemotherapy in advanced ALK-rearranged non-small-cell lung cancer (ASCEND-4): a randomised, open-label, phase 3 study‘, *The Lancet*. Elsevier, 389(10072), pp. 917–929.

Taagepera, S. *et al.* (1998) ‗Nuclear-cytoplasmic shuttling of C-ABL tyrosine kinase.‘, *Proceedings of the National Academy of Sciences of the United States of America*. National Academy of Sciences, 95(13), pp. 7457–62. Available at: http://www.ncbi.nlm.nih.gov/pubmed/9636171 (Accessed: 4 April 2018).

Tarrant, M. K. and Cole, P. A. (2009) ‗The chemical biology of protein phosphorylation.‘, *Annual review of biochemistry*. NIH Public Access, 78, pp. 797–825. doi: 10.1146/annurev.biochem.78.070907.103047.

Tokarski, J. S. *et al.* (2006) ‗The structure of Dasatinib (BMS-354825) bound to activated ABL kinase domain elucidates its inhibitory activity against imatinib-resistant ABL mutants.‘, *Cancer research*, 66(11), pp. 5790–7. doi: 10.1158/0008-5472.CAN-05-4187.

Verheijen, R. B. *et al.* (2017) ‗Clinical Pharmacokinetics and Pharmacodynamics of Pazopanib: Towards Optimized Dosing‘, *Clinical Pharmacokinetics*, 56(9), pp. 987–997. doi: 10.1007/s40262-017-0510-z.

Vihinen, M. *et al.* (1994) ‗Structural basis for pleckstrin homology domain mutations in X-linked agammaglobulinemia‘, *Biochemistry* , 91, pp. 12803–12807. Available at: https://www.pnas.org/content/pnas/91/26/12803.full.pdf (Accessed: 1 August 2019).

Vihinen, M. *et al.* (1996) ‗BTKbase, mutation database for X-linked agammaglobulinemia (XLA).‘, *Nucleic acids research*. Oxford University Press, 24(1), pp. 160–5. Available at: http://www.ncbi.nlm.nih.gov/pubmed/8594569 (Accessed: 3 April 2018).

Vlahovic, G. and Crawford, J. (2003) ‗Activation of Tyrosine Kinases in Cancer‘, *The Oncologist*. AlphaMed Press, 8(6), pp. 531–538. doi: 10.1634/THEONCOLOGIST.8-6-531.

Webersinke, G. (2016) ‗Molecular pathogenesis of chronic myeloid leukemia‘, *memo - Magazine of European Medical Oncology*. Springer Vienna, 9(4), pp. 163–167. doi: 10.1007/s12254-016-0294-0.

Wetzler, M. *et al.* (1993) ‗Subcellular localization of Bcr, Abl, and Bcr-Abl proteins in normal and leukemic cells and correlation of expression with myeloid differentiation.‘, *Journal of Clinical Investigation*, 92(4), pp. 1925–1939. doi: 10.1172/JCI116786.

Wong, S. and Witte, O. N. (2004) ‗The BCR-ABL story: bench to bedside and back.‘, *Annual review of immunology*, 22(1), pp. 247–306. doi: 10.1146/annurev.immunol.22.012703.104753.

Wu, P., Nielsen, T. E. and Clausen, M. H. (2015) ‗FDA-approved small-molecule kinase inhibitors.‘, *Trends in pharmacological sciences*, 36(7), pp. 422–39. doi: 10.1016/j.tips.2015.04.005.

Wylie, A. *et al.* (2014) ‗ABL001, a Potent Allosteric Inhibitor of BCR-ABL, Prevents Emergence of Resistant Disease When Administered in Combination with Nilotinib in an in Vivo Murine Model of Chronic Myeloid Leukemia‘, *Blood*, 124, pp. 398–398. Available at: http://www.bloodjournal.org/content/124/21/398.

Wylie, A. A. *et al.* (2017) ‗The allosteric inhibitor ABL001 enables dual targeting of BCR?ABL1', *Nature*, 543(7647), pp. 733–737. doi: 10.1038/nature21702.

Zhang, J. *et al.* (2010) ‗Targeting Bcr-Abl by combining allosteric with ATP-binding-site inhibitors.', *Nature*, 463(7280), pp. 501–6. doi: 10.1038/nature08675.

Zhang, S. and Yu, D. (2012) ‗Targeting Src family kinases in anti-cancer therapies: turning promise into triumph.', *Trends in pharmacological sciences*. NIH Public Access, 33(3), pp. 122–8. doi: 10.1016/j.tips.2011.11.002.

Zimmermann, J. *et al.* (1997) ‗Potent and selective inhibitors of the Abl-kinase: phenylamino-pyrimidine (PAP) derivatives', *Bioorganic & Medicinal Chemistry Letters*, 7(2), pp. 187–192. doi: 10.1016/S0960-894X(96)00601-4.

Zuccotto, F. *et al.* (2010) ‗Through the ―Gatekeeper Door": Exploiting the Active Kinase Conformation', *J. Med. Chem*, 53, pp. 2681–2694. doi: 10.1021/jm901443h.

FSC
www.fsc.org
MIX
Papier | Fördert
gute Waldnutzung
FSC® C083411

Kolibri 360 GmbH
Ferdinand-Jühlke-Straße 7
99095 Erfurt, Deutschland
produktsicherheit@kolibri360.de

Druck:
CPI Druckdienstleistungen GmbH
im Auftrag der
Kolibri 360 GmbH
Ferdinand-Jühlke-Str. 7
99095 Erfurt